THE POETRY OF MOSCOVIUM

The Poetry of Moscovium

Walter the Educator

Silent King Books

SILENT KING BOOKS

SKB

Copyright © 2024 by Walter the Educator

All rights reserved. No part of this book may be reproduced in any manner whatsoever without written permission except in the case of brief quotations embodied in critical articles and reviews.

First Printing, 2024

Disclaimer
This book is a literary work; poems are not about specific persons, locations, situations, and/or circumstances unless mentioned in a historical context. This book is for entertainment and informational purposes only. The author and publisher offer this information without warranties expressed or implied. No matter the grounds, neither the author nor the publisher will be accountable for any losses, injuries, or other damages caused by the reader's use of this book. The use of this book acknowledges an understanding and acceptance of this disclaimer.

"Earning a degree in chemistry changed my life!"
— Walter the Educator

dedicated to all the chemistry lovers, like myself, across the world

MOSCOVIUM

In the depths of the periodic table's maze,

MOSCOVIUM

Where elements dance in an atomic haze,

MOSCOVIUM

Lies a mystery, a shimmering delight,

MOSCOVIUM

Moscovium's realm, in the cosmic night.

MOSCOVIUM

Unveiled by science, in the fusion's glow,

MOSCOVIUM

Nuclei collide, in a stellar show,

MOSCOVIUM

Synthesized, crafted, in the lab's embrace,

MOSCOVIUM

Moscovium emerges, a fleeting grace.

MOSCOVIUM

Number one fifteen, its atomic decree,

MOSCOVIUM

A fleeting existence, for eyes to see,

MOSCOVIUM

Born from the union of elements rare,

MOSCOVIUM

Moscovium's essence, a cosmic affair.

MOSCOVIUM

With electrons swirling in quantum dance,

MOSCOVIUM

Moscovium's secrets, they seem to prance,

MOSCOVIUM

An island of wonder, in the periodic sea,

MOSCOVIUM

Beyond the reach of traditional chemistry.

MOSCOVIUM

Yet in its brief existence, it leaves a mark,

MOSCOVIUM

A testament to humanity's spark,

MOSCOVIUM

In the quest for knowledge, we strive to see,

MOSCOVIUM

The universe's truths, in its vast decree.

MOSCOVIUM

Moscovium, named for Russia's grandeur,

MOSCOVIUM

A tribute to science's enduring ardor,

MOSCOVIUM

In the land of Dmitri Mendeleev's design,

MOSCOVIUM

It joins the chorus of elements divine.

MOSCOVIUM

But what mysteries lie in its atomic core?

MOSCOVIUM

What secrets does it hold, forevermore?

MOSCOVIUM

Perhaps in its atoms, the universe sings,

MOSCOVIUM

Of the boundless wonders that chaos brings.

MOSCOVIUM

In laboratories hushed, with precision's art,

MOSCOVIUM

Scientists probe Moscovium's heart,

MOSCOVIUM

Unraveling mysteries, one by one,

MOSCOVIUM

In the quest to understand what's begun.

MOSCOVIUM

Its fleeting nature, a reminder profound,

MOSCOVIUM

That all in the universe, cycles bound,

MOSCOVIUM

From stars that burn bright, to elements rare,

MOSCOVIUM

Life's ephemeral dance, beyond compare.
MOSCOVIUM

So let us marvel at Moscovium's tale,

MOSCOVIUM

In the grand tapestry of creation's kale,

MOSCOVIUM

A symbol of humanity's ceaseless quest,

MOSCOVIUM

For truth and knowledge, we are blessed.

MOSCOVIUM

In the depths of the periodic table's maze,

MOSCOVIUM

Where elements dance in an atomic haze,

MOSCOVIUM

Moscovium stands, a shimmering light,

MOSCOVIUM

Guiding us onward, through the cosmic night.

MOSCOVIUM

ABOUT THE CREATOR

Walter the Educator is one of the pseudonyms for Walter Anderson. Formally educated in Chemistry, Business, and Education, he is an educator, an author, a diverse entrepreneur, and he is the son of a disabled war veteran. "Walter the Educator" shares his time between educating and creating. He holds interests and owns several creative projects that entertain, enlighten, enhance, and educate, hoping to inspire and motivate you.

Follow, find new works, and stay up to date with Walter the Educator™ at WaltertheEducator.com

www.ingramcontent.com/pod-product-compliance
Lightning Source LLC
LaVergne TN
LVHW010619070526
838199LV00063BA/5204